Plants

Reproduction in Flowering Plants

by Kate Boehm Jerome

Table of Contents

DEVELOP LANGUAGE

Some flowers open wide on warm, sunny days. The flowers are a good source of food for bees, butterflies, and other animals.

The animals help the plants, too. Animals can carry a plant's **pollen** to other plants. This helps plants make **seeds**.

Discuss the plants and animals you see on these pages. Ask and answer questions like these:

How does a hummingbird's long beak help it get food from plants?

How do you think honeybees move pollen?

How might a flower's bright color help it attract animals?

pollen – tiny grains that contain male sex cells
seed – a plant part that can grow into a new plant

The flower of a peanut cactus is bright red.

hummingbird
honeybee
pollen

CHAPTER 1

Flowering Plants

Many people enjoy flowers. They grow flowering plants in their yards and put flowers in vases to make their homes more beautiful.

But flowers aren't just for making our world a prettier place. They are a very important part of many seed-making plants.

There are two groups of plants that make seeds. **Angiosperms** are plants that produce flowers.

Angiosperms are the biggest group of seed plants. In fact, there are more flowering plants on Earth than any other kind.

Gymnosperms, the other group of seed-making plants, do not make flowers. Most gymnosperms produce their seeds on cones.

angiosperms – seed plants that produce flowers

gymnosperms – seed plants that do not produce flowers

This flowering cherry tree is an angiosperm.

Both gymnosperms and angiosperms are **vascular plants**. This means they have special tube-like tissues called **xylem** and **phloem**.

Xylem moves water and minerals from the soil through the plants. Phloem moves food through the plants. Since all the parts of vascular plants can get the **nutrients** they need, some vascular plants can grow very tall.

vascular plants – plants that have special tissues that move water, minerals, and food through the plants

xylem – the tissue in vascular plants that moves water and minerals from the soil through the plant

phloem – the tissue in vascular plants that moves food through the plant

nutrients – things that plants need to survive and grow

By the Way...

Not all plants are vascular plants. Liverworts are one example of a nonvascular plant. These plants do not have xylem and phloem. They cannot grow as tall as vascular plants.

This pine tree is a gymnosperm.

KEY IDEA Flowering plants are angiosperms.

Roots, Stems, and Leaves

Most flowering plants have roots, stems, and leaves. These parts help keep a plant alive.

Roots anchor, or hold, a plant in place. As roots grow, they also help keep soil in place around the plant.

Roots take in water and minerals from the soil. The plant needs these to grow and flower.

Stems support the plant. Like roots, they also contain xylem and phloem pathways. They allow water, minerals, and food to move between the roots and leaves.

Leaves are the plant parts in which most plants make food. All food-making leaves contain **chlorophyll.** Chlorophyll gives leaves their green color. It also allows plants to use energy from sunlight, or light energy, to make food. Without chlorophyll, the food-making process cannot begin.

chlorophyll – allows plants to use light energy to make food

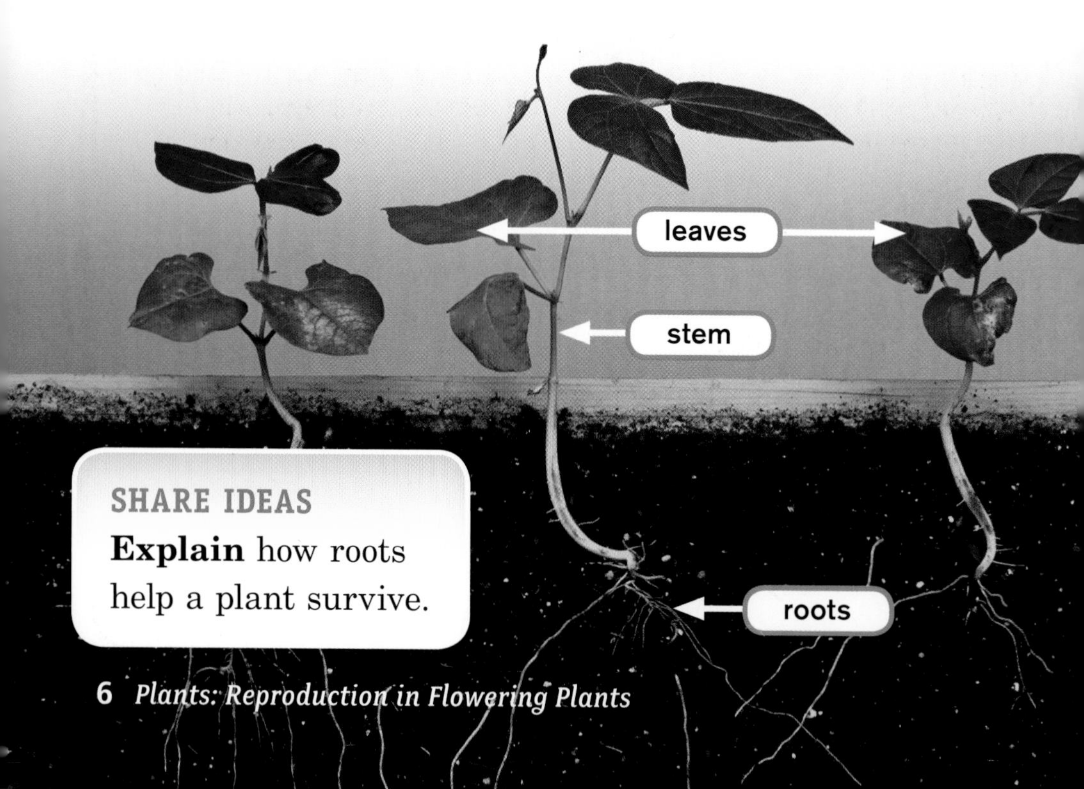

SHARE IDEAS

Explain how roots help a plant survive.

Photosynthesis

Flowering plants make their own food in a process called **photosynthesis.** They need light energy, water, and **carbon dioxide** for the process to begin.

During photosynthesis, plants use light energy to change water and carbon dioxide into sugar and **oxygen.**

The sugar is stored as food for the plant. Most of the oxygen is released into the air we breathe.

photosynthesis – the process in which plants use light energy to make food

carbon dioxide – a gas used during photosynthesis

oxygen – a gas produced during photosynthesis

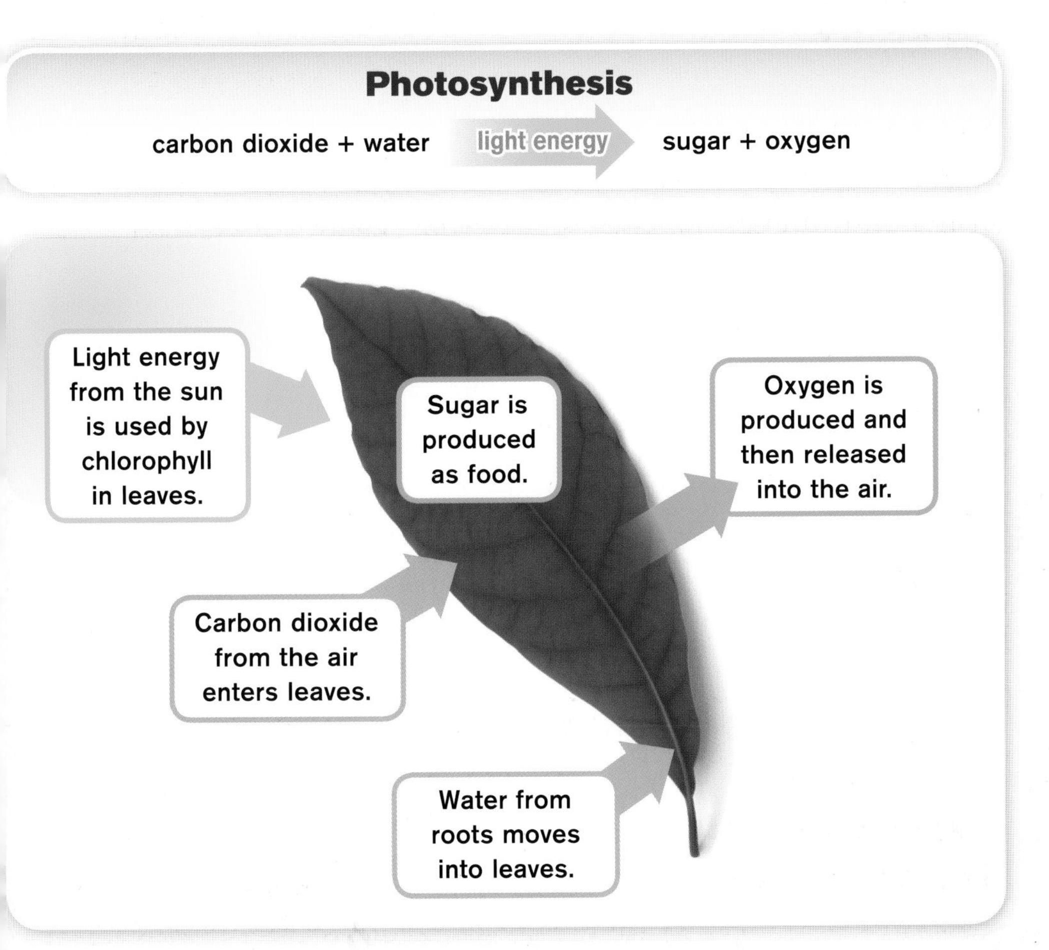

Respiration

Plants need energy to stay alive. During **respiration**, plants use oxygen to break down the stored sugar they made during photosynthesis.

Respiration releases energy that plant cells need for life processes. It also produces carbon dioxide and water.

The energy-releasing process of respiration does not need light. This means it can happen during the day or night.

respiration – the energy-releasing process in plants

KEY IDEAS Photosynthesis is the process in which plants use energy from sunlight to make food. Respiration is the process in which plants release the energy stored in food.

▼ **Flowering plants get energy by breaking down stored food.**

Photosynthesis	Respiration
• This process occurs only in cells with chlorophyll. • Energy is stored in sugar. • Carbon dioxide is used. • Oxygen is released.	• This process can occur in all cells. • Energy is released from sugar. • Carbon dioxide is produced. • Oxygen is used.

YOUR TURN

Look at the flowers on this page. Can you think of a way to sort them into two groups?

Can you think of other ways that scientists might group different flowers?

Describe how they might do it.

MAKE CONNECTIONS

Different plants grow in different places. Think about the flowering plants that grow where you live. Describe the kinds of flowers you've seen.

USE THE LANGUAGE OF SCIENCE

What happens in plants during photosynthesis and respiration?

During photosynthesis, plants make food. They release energy from stored food during respiration.

CHAPTER 2

Sexual Reproduction in Flowering Plants

There are many different kinds of flowers. Even corn plants have flowers.

All flowers have the same job. Through **sexual reproduction,** they help the plant make seeds. The seeds can then grow into new plants.

Sexual reproduction requires male and female sex cells to join together. Sex cells have material needed to form a new living thing. The flowers of most plants have both male and female parts. However, some plants have flowers with only male parts or only female parts.

sexual reproduction – a process that requires the combination of male and female sex cells to make a new living thing

These are the male flowers on a corn plant.

Most flowers have four main parts, the **sepals**, **petals**, **stamens**, and **pistil**.

The green, leaf-like sepals cover and protect the unopened flower, or bud.

As the flower grows, the sepals open and the flower petals begin to show. Petals are the parts of the flower that often have a lot of color.

The stamens, which are just inside the petals, are the male parts of the flower. Stamens produce tiny grains called pollen. Pollen contains the male sex cells needed to make seeds.

The pistil, or the female part, is in the very middle of the flower. The pistil produces the female sex cells needed to make seeds.

sepals – the parts of a flower that protect the bud

petals – usually colorful parts of a flower

stamens – male parts of a flower

pistil – a female part of a flower

Explore Language

LATIN WORD ROOTS

pollen

pollen or *pollis* = fine flour

Parts of a Flower

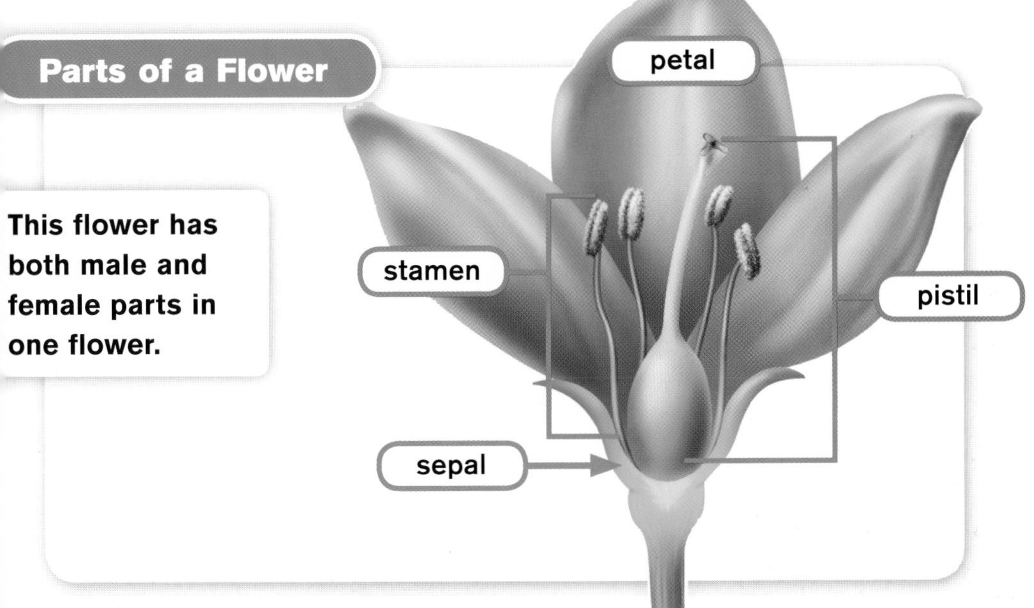

This flower has both male and female parts in one flower.

Pollination

Pollen from a flower's stamen needs to be moved to a pistil before a seed can form. This happens in a process called **pollination.** Sometimes pollen is moved to a pistil in the same flower or to another flower on the same plant.

Most often, however, pollen is moved from one plant to another.

Wind can blow pollen from plant to plant. Animals, such as butterflies and bees, can also move pollen. Pollen can stick to them when they land on a flower. When they visit another flower, the pollen can rub off on the flower's pistil.

▲ **The bee carries pollen from one flower to the next.**

pollination – the process in which pollen is moved from a stamen to a pistil

KEY IDEA Pollination occurs when a flower's pollen is moved from a stamen to a pistil.

BY THE WAY...

Some people are allergic to pollen. Pollen can make them sneeze and cause their eyes to itch.

Fertilization

The pistil is the female plant part in which a seed is made. When pollen lands on the pistil, **fertilization** can occur.

For fertilization to happen, male sex cells from pollen have to move through the different parts of the pistil.

Fertilization happens when the male and female sex cells join together inside the pistil. After fertilization, a seed can develop. The seed can grow into a new plant.

fertilization – the process in which male and female sex cells join to form a new living thing

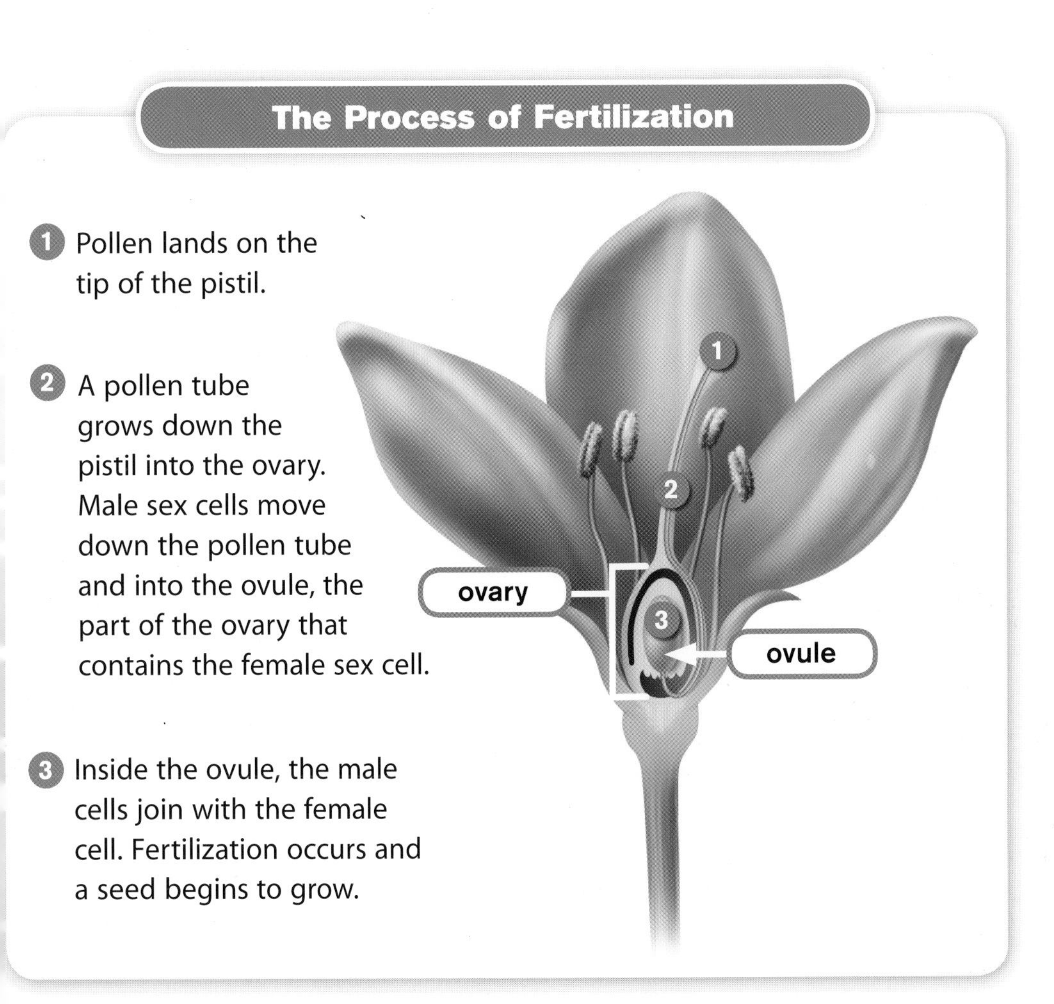

Fruit and Seeds

Most parts of the flower are no longer needed after fertilization, so petals fall to the ground.

But the ovary slowly changes into a **fruit** that protects the seeds while they grow.

When the fruit is ripe, the seeds are ready to grow into new plants. Sometimes a fruit just falls on the ground and decays, or rots. The seeds inside are left on the ground to grow.

Other times, fruits and seeds are carried by wind or animals to other places. If seeds land in places that have good soil and water, they have a chance to grow into new flowering plants.

fruit – a covering that protects the seeds of flowering plants

The wind can blow dandelion seeds to different places.

The orange fruit protects the seeds inside.

orange

KEY IDEAS Fertilization occurs when male sex cells join female sex cells to form new living things.

YOUR TURN

OBSERVE

Bees are not the only animals that help with pollination. Butterflies can, too. Look at the picture. Explain how a butterfly can help pollinate a flower.

MAKE CONNECTIONS

Most plants that grow in the hot desert have only a short time to reproduce after it rains. In that short time, many desert plants produce large numbers of flowers. Why do you think they do this?

STRATEGY FOCUS

Make Connections

Look back at this chapter. What connections can you make? Make a chart like this one. Write down your connections.

The text says . . . or The picture shows . . .	This reminds me of . . .	It helps me understand that . . .
(p. 14) If seeds land in places that have good soil and water...	new wildflowers that grow each year by my house	seeds are carried by the wind and by animals

CHAPTER 3

Asexual Reproduction in Flowering Plants

Did you know you can grow a whole new plant from a single sweet potato? But a sweet potato is not a seed. So how does the new plant grow?

Some angiosperms can use **asexual reproduction** to grow new plants. In asexual reproduction, fertilization does not occur. Instead, new plants grow from a plant part.

The sweet potato is the root of a flowering plant. New plants can grow from the root.

asexual reproduction – a process that produces new plants without fertilization

There are many different ways that asexual reproduction can occur. New plants can grow from roots, like sweet potatoes, or from leaves and stems.

African violet plants can grow from leaves. If a leaf is cut off and put in water or soil, a new plant grows.

Strawberry plants can grow from stems called **runners** that grow on top of the soil. When the runner touches the ground, it grows roots and a new strawberry plant can grow.

runners – stems that grow on top of the soil and produce roots

New strawberry plants grow from runners that send roots into the soil.

A Useful Way to Grow

Asexual reproduction is useful for people who grow plants. It can be used to produce new plants that are identical, or exactly the same.

For example, new plants grown from African violet leaves will be the same as the plant the leaves came from. The new plants will have flowers that are the same color and size. If new plants are grown from seeds, they will not be identical.

Asexual reproduction can also produce new plants more quickly. This can be important for people who sell plants for a living.

No matter how new plants grow, one thing is certain. Plants are an important part of our world. They produce the oxygen and food that living things need. Without plants, life on earth could not survive.

KEY IDEAS Asexual reproduction does not require fertilization. Asexual reproduction makes new plants that are identical.

Asexual reproduction can produce plants faster.

INFER

Look at each of the pictures. Infer whether sexual or asexual reproduction is occurring.

MAKE CONNECTIONS

Seedless grapes are easier to eat. Do you think these plants are grown asexually or sexually? Explain why.

EXPAND VOCABULARY

During respiration, plants and animals give off carbon dioxide. Many animals get rid of this carbon dioxide when they breathe out. *Respiration* comes from the Latin word *respirare*, which means "to breathe." Find other words that come from *respirare*. Explain how each word is related to breathing.

Horticulturists: What Do They Do?

Horticulturists are scientists who find ways to grow healthier plants. Some do research. Others work with businesses that grow plants. Some horticulturists like to teach. But no matter what job they do, all horticulturists love plants!

Places a horticulturist might work	• research laboratory • garden center • university • zoo or botanical garden
Education a horticulturist needs	• most horticulturists have a college degree
Courses you should study if you want to be a horticulturist	• chemistry • botany • biology

CAREER PROFILE

KATIE COOK
Interpretive Horticulturist
Education Department
Callaway Gardens
Pine Mountain, Georgia

Language that Compares

'o compare two or more things, you can use special adjectives called omparatives and superlatives.

Adjective	Comparative	Superlative
tall big	taller bigger	tallest biggest

XAMPLE

The tree is **taller** than the daffodils.

The gladiolus is the **tallest** flower in the garden.

ome comparatives and superlatives don't follow the pattern:

bad, worse, worst

good, better, best

little, less, least

much, more, most

'age through this book. Talk about the flowers using comparatives nd superlatives.

Write to Compare

Compare an apple tree, a strawberry plant, and a sweet potato plant. Use comparatives and superlatives.

Words You Can Use		
Adjective	**Comparative**	**Superlative**
tall large short small fragrant	taller larger shorter smaller more fragrant	the tallest the largest the shortest the smallest the most fragrant

PLANTS TO THE RESCUE!

Echinacea
may help fight infection.

Cayenne
may help pain in joints.

Some flowering plants are used as medicines. In the past, groups of Native Americans used coneflower plants to treat burns and insect bites. A substance in the foxglove flower was once used to treat heart conditions.

CAUTION

Never eat any part of a plant in the wild. Some plants are very harmful, even deadly!

Look at the poster.

- What are some plants that have been used as medicines?
- How did people think these plants could help their health?

Key Words

ngiosperm (angiosperms) a seed lant that produces flowers
rose bush is an **angiosperm**.

sexual reproduction a process hat produces identical new plants ithout fertilization
n African violet leaf can grow a new lant through **asexual reproduction**.

hlorophyll the substance that ives plants their green color and llows plants to use energy from he sun to make food
green leaf contains **chlorophyll**.

ertilization the process in which nale and female sex cells join to orm a new living thing
ertilization happens when male and emale sex cells join together.

ruit (fruits) a covering that rotects the seeds of flowering lants
n apple is a **fruit** with many seeds.

ymnosperm (gymnosperms) seed plant that does not roduce flowers
pine tree is a **gymnosperm**.

hloem the tissue in vascular lants that moves food
hloem carries food from the leaves rough the plant.

hotosynthesis the process in hich plants use energy from unlight to make food
reen plants make food through **hotosynthesis**.

pistil a female part of a flower
The **pistil** of a flower contains the female sex cells.

pollen tiny grains that contain male sex cells
Pollen must reach the pistil of a flower for fertilization to occur.

pollination the process in which pollen is moved from a stamen to a pistil
Bees and butterflies often take part in the process of **pollination**.

respiration the energy-releasing process in plants
Plants release the energy stored in food during **respiration**.

sexual reproduction a process in which male and female sex cells join to form a new living thing
Many plants make new plants through **sexual reproduction**.

stamen (stamens) a male part of a flower
A **stamen** makes pollen that is needed for sexual reproduction.

vascular plant (vascular plants) a plant that has special tissues to move nutrients through the plant
A **vascular plant** can grow taller than a nonvascular plant.

xylem the tissue in vascular plants that moves water and minerals from the soil
Xylem moves water and minerals from the roots through the plant.

Index

MILLMARK EDUCATION CORPORATION
Ericka Markman, President and CEO; Karen Peratt, VP, Editorial Director; Rachel L. Moir, Director, Operations and Production; Mary Ann Mortellaro, Science Editor; Amy Sarver, Series Editor; Betsy Carpenter, Editor; Guadalupe Lopez, Writer; Kris Hanneman and Pictures Unlimited, Photo Research

PROGRAM AUTHORS
Mary Hawley; Program Author, Instructional Design
Kate Boehm Jerome; Program Author, Science

BOOK DESIGN Steve Curtis Design

CONTENT REVIEWER
Nikki L. Hanegan, PhD, Brigham Young University, Provo, UT

PROGRAM ADVISORS
Scott K. Baker, EdD, Pacific Institutes for Research, Eugene, OR
Carla C. Johnson, EdD, University of Toledo, Toledo, OH
Donna Ogle, EdD, National-Louis University, Chicago, IL
Betty Ansin Smallwood, PhD, Center for Applied Linguistics, Washington, DC
Gail Thompson, PhD, Claremont Graduate University, Claremont, CA
Emma Violand-Sánchez, EdD, Arlington Public Schools, Arlington, VA (retired)

PHOTO CREDITS Cover © JoLin/Shutterstock; 1a © Elena Elisseeva/Shutterstock; 2-3 © Rob Huntley/Shutterstock; 2a © Charles W. Melton; 3a © William Leaman/Alamy; 3b © Sasha Radosavljevich/Shutterstock; 4a and 15b © Cornelia Doerr/age fotostock; 5a © Ed Reschke/Peter Arnold, Inc.; 5b © Wendy Nero/Shutterstock; 6a © S.J. Krasemann/Peter Arnold, Inc.; 7a © Olga Shelego/Shutterstock; 8a © Jean Carter/age fotostock; 9a © Rolf Klebsattel/Shutterstock; 9b © Khramtsova Tatyana/Shutterstock; 9c © Kelly McAdam/Shutterstock; 9d and 9e Lloyd Wolf for Millmark Education; 10a © Michael J. Hipple/age fotostock; 11a and 13a Illustrations by Joel and Sharon Harris; 12a © Hans Pfletschinger/Peter Arnold, Inc.; 12b © Mixa/Getty Images; 14a Pablo Galán Cela/age fotostock; 14b © Photodisc/Punchstock; 15a © Goran Kapor/Shutterstock; 16a © Robbert Koene/Getty Images; 17a © Evan Sklar/Botanica/JupiterImages; 17b © Henry Beeker/age fotostock; 18a © NouN/Garden Picture Library; 19a © Dynamic Graphics Group/Creatas/Alamy; 1 © AJPhoto/Photo Researchers, Inc; 19c © Thomas Photography LLC/Alamy; 20a Courtesy Katie Cook; 20b © Westend61/Punchstock; 22a © Yanik Chauvin/Shutterstock; 22b © Mauritius/Photolibrary; 24 © Darrell Young/Shutterstock

Published by Millmark Education Corporation
7272 Wisconsin Avenue, Suite 300
Bethesda, MD 20814

ISBN-13: 978-1-4334-0045-2
ISBN-10: 1-4334-0045-6

Printed in the USA

10 9 8 7 6 5 4 3 2 1